"I haven't read this book yet, but the clarity and candor implied in the title alone make me think it could be the greatest-ever book of poems about barns."

– **DAVE EGGERS**, AUTHOR OF *THE CAPTAIN AND THE GLORY*

"A fun exploration of the power of arbitrary constraints, with Brautiganesque, perverse humor. I now know as little about barns as I did before."

– **STEPHIN MERRITT** OF THE MAGNETIC FIELDS

"*50 Barn Poems* seems like it's making fun of barns, or poems, or poets, or books. I like that about it."

– **ELIZABETH ELLEN**, AUTHOR OF *ELIZABETH ELLEN* AND *PERSONIA*

"Zesty. This book zings. Ever surprising, ever brilliant, Zac Smith is the David Shrigley of poetry."

– **Lars Iyer**, author of *Spurious* and *Wittgenstein Jr*

"*50 Barn Poems* is one of the most fun collections I've ever read... down-to-earth poems that are simply delightful, often beautiful, and occasionally downright brutal – in a heavy metal sort of way."

– **BENJAMIN DEVOS**, AUTHOR OF *HUMAN FISH* AND *THE BAR IS LOW*

"Unsentimental but somehow moving, funny without being cynical, these poems *are* the barn. They're everywhere and nowhere, utilitarian and excessive, like all beautiful structures."

– **LINDSAY LERMAN**, AUTHOR OF *I'M FROM NOWHERE*

"Zac Smith had a lot of fun writing these poems. I got the feeling that he drank a lot of coffee and let his imagination lead him. I think that's beautiful. He is winning the war against adulthood. Zac Smith will never sell out."

– **BLAKE MIDDLETON**, AUTHOR OF *COLLEGE NOVEL*

"With *50 Barn Poems*, Zac Smith harvests a bountiful collection that will sustain you with infinite laughter, surreal revelations and poignant truths."

– **MIKE ANDRELCZYK**, AUTHOR OF *THE IGUANA GREEN CITY AND OTHER POEMS*

"They said it couldn't be done, but Zac Smith has written fifty wondrously deranged and highly entertaining poems about barns. Maybe it's like *Animal Farm* for degenerates, or maybe it's unlike anything you've ever read before. Maybe it's beautiful."

—**BRIAN ALAN ELLIS**, AUTHOR OF *SAD LAUGHTER*

"All the facets of life can be found inside a barn if only one were to look through Zac's glasses."

- **CAVIN B. GONZALEZ**, AUTHOR OF *I COULD BE YOUR NEIGHBOR, ISN'T THAT HORRIFYING?*

"50 poems about the earth and what it can offer and how to get there. The earth is a poem about being, and being is profane and silly. These poems are a tour of celebrity barns. They understand that nature is glamorous and that glamour is actually the small things that people do when the world (or maybe God) isn't watching."

- **DANIEL BAILEY**, AUTHOR OF *THE DRUNK SONNETS* AND *GATHER ME*

"These poems fall across America with their heavy beams and flaking paint. Zac Smith has captured absurdity with empathy and humour in a collection that feels wholly contemporary."

- **GIACOMO POPE**, FOUNDER OF NEUTRAL SPACES AND AUTHOR OF *50 BARN BLURBS*

50 BARN POEMS

ZAC SMITH

CL◀SH

For Beary Sweet

CONTENTS

BARN POEM 1

not gonna look up anything about barns
these are barn memories
barn feelings
barn impressions

these are the barns in my head
big and red
or maybe grey
and dilapidated

these are barn poems
welcome

welcome to the barn,
motherfucker

BARN POEM 2

barn on the side of the road
built way too close to the shoulder

every three months a car hits it
every three months the sheriff makes the same joke

something about a drunk driver hitting the broadside of a barn
it's exactly as funny every time, too

BARN POEM 3

i ask my friend Murat about barns
"what do you know or think about barns?" i ask
he says:
"idk man"
"never thought about barns"
"really"
" 🙂 "

and he says:
"there is only so much i can think about"

BARN POEM 4

james had a ping pong table in the barn
on the open-air second floor

we put on shorts and tried to hit with a spin
every other ball landing in the cow shit

or the horse shit
down below

BARN POEM 5

rooster on top of the barn
holy shit
how'd he get up there?

he cock-a-doodle-doos all afternoon
afraid to jump
afraid to stay

BARN POEM 6

1.
oh man
is that a barn?

never seen one in real life before
nice

how many horses you fit in that thing?
dang, probably a lot

2.
woah haha that dude's really in there
in the barn, haha

hahaha wtf
look at that

why is he in there
oh shit lmaoo

3.
dude's still in the fuckin barn
hahahaha

goddamn dude
get outta there

is that his barn?
can't be his barn, right?

4.
dude's in the barn again
lmaoo haha

get outta there dude!
cow's gonna get you!

haha shit, what a nut
fuckin' love that guy

BARN POEM 7

trying to remember other barns
no one barn but a blur of many barns
thousands, probably, of anonymous monoliths
all haunting the fields along which i've driven
or been driven

mostly iowa and ohio and michigan
but new york, too. pennsylvania, massachusetts

i drove from upstate new york to seattle once
probably saw at least one barn in every state on that route –
minnesota, south dakota, north dakota, idaho, montana

there was an old farmhouse on a mountain
nothing else around, nothing on the radio

fear lived inside me then, being so distant from everything
anything could go wrong, it felt like
and the old farmhouse knew this –
it knew how small i felt
how small i was
weak and insignificant

a speck in the nowhere

or maybe it was a barn
i dunno
it's hard to remember
let's say it was a barn
we have that power

we can turn anything
into a barn

BARN POEM 8

tornado boy's coming
touch down! touch down!

oh shit oh no
there goes the barn!

look at that
that's insane

holy shit
holy fucking shit

BARN POEM 9

barn full of confetti
don't open that door!

hey!

ah shit
fuck dude

this sucks

look at all this
it's all damp, too

ugh wtf

BARN POEM 10

we all belong to a barn somewhere –
there is no escape

BARN POEM 11

1.
hot bales of hay
rip one open and see the steam

james says they can get really hot
even burn down the barn

it could be a perfect crime
no one would be to blame

2.
volunteering at the organic farm for the summer
i don't remember the barns

i just remember the hot, smoldering bales of hay
and the dogs that ran around in that special dog way

BARN POEM 12

barn on wheels
cruisin' down the highway

righteous cliffs
and sandy shores

hairpin turn!
oh shit!

barn in the air
a slow twirl toward the sea

BARN POEM 13

1.
neverending cemetery
with tiny barns
instead of headstones

at night you might see
one of the barns
hosts a small shining light

2.
barn with no doors –
trapped in a barn box

infinite funeral
upon the hay

3.
underneath the mountains

lurk five million barns
and inside each barn
is five
million
smaller
barns

4.
i open the barn
and it is full of snakes

BARN POEM 14

welcome to the dairy
this barn has special flooring
fifty cows standing in their own shit

the cows are like family
hose them off
so they don't start to smell

and the guatemalan family?
they spray the hoses
and they are also like family

BARN POEM 15

1.
the secret path through the little woods
the long way to james's barn

i can't remember what it smelled like
but i remember the soft dirt

we never walked the path enough
to harden it underfoot

2.
couple cats in here
three, maybe four

james says they're barn cats
and that they all have aids

if you get feline aids
you become a barn cat

3.
catnip hung to dry in the barn
dust flitting in the light

i watch a cat leap from the top of the haystack
onto the back of the cow
and disappear

BARN POEM 16

i try to think about what my dad said once
one trip through illinois
or indiana, or somewhere else

something about taxes –
big old barn gutted, broken
leaning to one side –

it obscures endless fields of soybeans,
a mansion,
and a four-car garage

BARN POEM 17

don't give birth in a barn
what the fuck is wrong with you

you're so close to letting your baby
be born into the beautiful golden sunlight

the big blue sky
could be the first thing your baby sees

it's right there
right out the barn door

but nah nah
a fuckin' barn, huh?

cool, no no it's cool
cool barn

sounds great
go for it

fuckin'
barn birth

yeah no cool
i love it

BARN POEM 18

1.
barn full of gold bars
why not?

fuck it

2.
you can't hack a barn, dummy
no computers in there!

we're safe (bring in the gold bars!)

3.
barn built out of a thick titanium alloy
child's drawing, withstanding missiles and bombs

you can leave your gold here. it's safe with me

BARN POEM 19

i forget the name of the goat with the long,
long tongue and the cleft palette
oozing snot 24/7

it must have slept in the barn
but i only remember it prancing around the field, screaming

maybe it never slept
maybe it only pranced and screamed

BARN POEM 20

barnyard with no fence
where does it end?

fuck man i dunno
probably at the next barn

haha fuck
endless barnyards

BARN POEM 21

here comes the barn
rattlin' rattlin' barn

watch out!
oh damn!

hitting the crest of the old country road
and getting some air

oh my god
look at that

sick fucking barn stunts
sick-ass barn stunt dude

hell yeah dude
hell yeah

do it again, man
some sick fuckin' barn stunts dude

BARN POEM 22

the heart is a barn
just open up and
see for yourself

things live there
but you gotta feed 'em
and shovel their shit

BARN POEM 23

Winter, time for the winter coat. Time for a big warm coat in the big cold world. His old pea coat with the big buttons, the big pockets. In the right-hand pocket, he feels something old and familiar. It is like a vague memory from childhood. A place from a dream. He holds it, prods it, looks for the meaning with touch alone. He tries to remember it, its strange shape, its give, its texture, but he can't place it, pulls it out, looks at it in the light. It is a hair tie. He sees that it is a hair tie, and bound around the stiff little pinched part is some hair. A small bundle of hair, wound up tight, brittle, frayed, wild. When was this from? When was this put here? He wonders. He can't remember. Hair ties were always everywhere. Under the pillows, behind the couch, in the cupboards, in the heating vents. They would pop up under rugs, in the yard, hanging from the back of a radiator. In the spring, he looked for them all. And he opened the windows, aired out the drawers, swept and dusted and scrubbed. He cleaned and cleaned and cleaned. Not just the hair ties, but the bobby pins, too. And the chapsticks, the soaps, the magazines, the tiny balled-up socks, the letters from the electric company, the letters from the bank, the credit card advertisements, the old postcards. Each with her name. Their address. The big empty house. He hasn't worn the coat since February. Then it was March, April, May. Time kept happening around him. Now it is November. Now it is November and he has her hair tie. He has her hair. And that's it, that's all, that's all he has. Oh, and he's in a barn. I forgot to mention that. That's what makes this a barn poem.

BARN POEM 24

ping pong ball skitters off my paddle
and james runs down the stairs, tosses it back up to me

he climbs up the side of the stable
deft, and with a small grunt

and i think:
"he's always been good at climbing"

i think:
"he's always been good at ping pong, too"

i look at the muscles in his arms and think:
"good arms for a barn"

i think:
"good barn arms"

BARN POEM 25

halfway there
halfway to barn town, babbyy

BARN POEM 26

imagining a guy posting an optical illusion to twitter
a seizure-inducing conflab of colors, stripes, and spheres
and he says,
"all these balls are the same color!"
he says,
"and that color is: BARN"

BARN POEM 27

barn on a skateboard
nollie kickflip to crooked grind

nails the landing
fuckin' natch

hell yeah dude
that's so fuckin' sick

but then the deck snaps
and the barn collapses

BARN POEM 28

1.
i don't know how the rain sounds from inside
a barn

2.
just wanna feel like a dog in a barn
with a cool breeze coming in through the open window

the sky would be blue, too
(feels important to clarify that)

3.
how a dog twitches and snores in her sleep
is how i want to be in the barn

how a cow sleeps, however that is
is how i want to be in the barn

how the barn cats sleep after hunting for mice
is how i want to be in the barn

i want to sleep in a barn, i guess, is the idea

BARN POEM 29

put a barn downtown somewhere you fucking cowards

bring the cows on home
bring the horses, too

let's have a fucking party

BARN POEM 30

a big, beautiful, blood-red barn
yeah, nice

it was painted with blood
took, like, a whole lotta blood

you wouldn't believe
just how much blood

BARN POEM 31

guy pulls off county road 15
takes off his driving gloves, stuff them in his back pocket

walks up the shoulder
(spurs clicking, of course)

he stands still
hands on his hips

smacks those manly, manly lips
"pretty good barn," he says

then he winks at the camera
;^)

BARN POEM 32

overthrow the government
i'm not fucking kidding here
tear it the fuck down

the barn awaits

BARN POEM 33

thinking about our beautiful shining future
humanity in repose, tranquil and serene:

o beautiful for halcyon barns
barn to shining barn

the intercontinental barnway
the transatlantic barnline

the great barn of china
the great barrier barn

the space shuttle, but, like,
it's a barn

hell yeah dude
that's beautiful

BARN POEM 34

a barn in the ocean
it creaks, old and weary

a hundred seagulls
instead of the solitary rooster

may it someday wash ashore
when you least expect it

BARN POEM 35

two barns hovering above the fruited plane
emitting a low drone
and glowing faintly

i think back on this often
maybe it is a dream
or maybe it is the future

BARN POEM 36 (3 HAIKU)

1.
broken social scene wrote that song about dreaming about me
and barns

2.
james put bss
in his myspace page interests
but left out
the barn

3.
out of all the barns, real or imagined, i still go back to
that barn

BARN POEM 37

ready to fucking kill someone
ready to slit some fucking throats

ominous barns on the hill
peaking down in the moonlight

they rattle around
like cows' bells

BARN POEM 38

life in a barn underwater
the roof pops and bubbles come out
like one of those clam shell things in a fish tank
but each time, some of my brain cells
float away
f o r e v e r

BARN POEM 39

the barn stands tall next to endless apple trees and fields of
grain
a svelte romanticization of the colonial west
and yet here i am, 39 poems deep

i am considering the barn
a pluck of wheat
perched precariously on my lip

BARN POEM 40

a barn instead of a coffin
moo moo, motherfuckers

BARN POEM 41

as children, we spend hours memorizing facts about exotic
animals, far-off places. planets, stars. quicksand. our parents
and teachers and caregivers teach us exotic escapism as a way
to interest us in lives outside of the very proximal. no teacher
wants their student to grow up to be a truck driver. so it's all
flamingos, elephants, toucans, polar bears. we forget all this
stuff as we grow up, but because we're so trained to ignore the
stuff around us, it's like we don't know anything when we're
older. the trees are boring, the squirrels are boring, the rivers
are boring. there are no elephants or flamingos, none of the
really fascinating stuff. so instead we look inward, we look
stuff up in books. we come up with ways to distract ourselves
as our imaginations die out. we learn that there is no one
notion of thinking, so we learn new ways to think. we learn
new ways of categorizing things. new ways of seeing the
boring world around us. someone says you can do this enough
and eventually you can get paid to do it, get paid to teach, get
paid to think about thinking. but barns are never boring, and
they never will be. don't forget that. it begins and ends with a
barn.

BARN POEM 42

when i die
don't bury me in a barn

that's not actually a thing
don't get it twisted

burn my corpse
as you would anyone else

but also, please
scatter my cremains

where a barn once stood

BARN POEM 43

barn stunts!
they're back!

grab a can of red bull!
and meet me by the barn!

wooooo!

it has been popping off like this
all
fucking
day
!!!

splash

wooooooooo!

BARN POEM 44

you can turn out the lights and you can open the windows and
you can throw out all your furniture and you can invite some
cows and goats and a horse over and you can get a couple
bales of hay and you can pretend that your little apartment is
actually a barn

BARN POEM 45

welcome to the jungle, baby
we got fun and games

i mean the barn
welcome to the barn, baby

etc. etc. etc.
it's basically a jungle

like, it'll fucking kill you
oh my god, it can absolutely kill you

don't be fooled
it's dangerous as fuck

i mean it's basically a jungle
might as well be the jungle

i call my barn "the jungle", for example

BARN POEM 46

i disassemble the house when he isn't around
piecemeal and in secret

i remove the outlet covers
first in the basement, then the pantry, then the hallways

he leaves for the day and i snap the casements off the windows
first one room, then another

i take steel wool to the paint in the dining room
one wall at a time, sweep up the specks of dried paint

slowly, slowly, slowly, i scrape off all the paint
then the next room, the baseboards, the wainscoting

one wall at a time
the hallway, the den, the kitchen

i sneak out onto the roof while he sleeps
i rip off shingles, two at a time

fold them until they crack, carry them inside

bury them in the trash under tissues, cellophane, coffee
grounds

i pull out nails, old ones, from which pictures used to hang
nails from the floorboards, too, and the cabinets, the
windowsills

i peel up the floorboards hidden under the overhang of the
kitchen cabinets
they snap easily over my knee

i scatter the splinters in the woods
where the wood belongs

i pry tiles out of the shower, throw them into the dense weeds
in the backyard
but otherwise i leave the weeds as they are – big and powerful,
full of secrets

and i start in on the bricks in the basement
the flaky mortar, the cement, the foundation

i will do this until he notices
i will do this until he asks me what is wrong

and then i will begin to repair the damage
or else i will move onto the barn

BARN POEM 47

rented u-haul on the two-lane country road winding through
upstate new york
we crest a hill accelerating just past another 35-mph zone

a single barn surveys its kingdom below:
it is all endless valleys
and smaller hills one could crest
with enough speed and purpose

BARN POEM 48

barn carving a wave instead of nestled on a hill
black surfboard and a lit cigarette

when i was a kid, i had a bowl cut. the barber called it a
surfer cut
so of course i thought it was radical

maybe i still do
(or maybe i don't)

but when i think about barns doing sick ocean stunts
none of them have bowl cuts

BARN POEM 49

no one has ever built a barn
they just sprout up

it's beautiful, man
how nature provides

BARN POEM 50

died in a barn
RIB
(rest in barn)

ACKNOWLEDGMENTS

This book only exists due to the unending support and kindness of the following wonderful people: master editor and friend extraordinaire Cavin B. Gonzalez, King Pope Giacomo Pope, Master Poet Mike Andrelczyk, Nick (poet champion) and Liz for being the hands-down best fans, Lyla, Luna, Stella, Lorenzo (Flat Pet City Crew), James & Julie & Robin & Brad & the Barn, Dr. Brent Woo, The Dave Eggers Fan Club, @a_neutral_space, Matthew Revert, Lindsay, Ben, Blake, Brian, Elizabeth, Claudia, Stephin, Lars, Dave (Eggers), my family, my extended family, all the editors & readers of all the journals I know and love and all the writers who write for them, and of course Leza and Christoph and everyone else at CLASH, obviously, you kidding me? But, ultimately, thank you most of all to Jessica: my beep, the critic I trust most, the reason I write. Thank you.

ABOUT THE AUTHOR

Author photo by Eric Quesada

Zac Smith lives in Boston, where he likes to walk his dogs.
This is his first book. Thank you.
Follow him on Twitter @ZacTheLinguist
zacsmith.net

CLASH

WE PUT THE LIT IN LITERARY

FOLLOW
TWITTER
IG
FB
@clashbooks

EMAIL
clashmediabooks@gmail.com